MATH IS EVERYWHERE!

Photos to inspire math talk

Grades K - 2

by Colleen Uscianowski, PhD

LUMINOUSLEARNING

10 9 8 7 6 5 4 3 2 1

Published by Luminous Learning Inc, New York
Orders: www.luminouslearning.com
info@luminouslearning.com

ISBN 978-1-939763-00-6

DEAR FELLOW MATH ENTHUSIAST,

Opportunities to notice and talk about math are everywhere!

Making math part of your daily routine helps students find the fun and the relevance in math. This can be especially helpful for students with disabilities or those with math anxiety. The photos you'll find in this book are examples of real world math, which help our students see math as more than a set of rules we memorize in school. Use the photos to show students how we can find math in our everyday world, such as outside in nature, around our neighborhood, in our home and classroom, and at the supermarket.

Here are some ways to use these photos in your classroom:

Ask questions to engage your students in math talk. Ask students open-ended questions about each photo, such as, "What do you notice? What math do you see?" Celebrate all of the different kinds of math that students can find. Some students may notice numbers, while others will see shapes or examples of measurement. This book also provides specific math questions you can ask about each photo. You can come up with some of your own questions, too! A great follow-up question to ask students after any of the questions in this book is, "How did you know?"

Turn these math photos into word problems. Have students write their own word problems and swap with a partner to solve. Encourage students to draw pictures or diagrams to show the math they see in the photos as they solve the word problems.

Encourage students to find their own examples of math. Give them cameras, tablets, or drawing tools that allow them to capture the math they encounter in everyday life. Take a walk inside the school or around the neighborhood. Have students take or draw pictures of the math they find. Post their photos up in the classroom and have a gallery walk to allow students to notice and comment on the math their classmates captured.

Check out the accompanying workbook: **MATH IS EVERYWHERE WORKBOOK**
Visit our YouTube channel to learn more about how to use this book:
youtube.com/user/Luminouslearning

How many groups of bagels are there?
How many bagels are in each group?
How many bagels are there in total?

How else could you separate this many bagels into groups?

If you ate half of the bagels, how many would be left?

ATLAS KITCHEN

What pattern do you see? If this pattern continued, what would come next?

Can you make this same type of pattern using numbers?

Can you make this same type of pattern using body movements?

What shapes do you see?
How are these shapes different?
How are these shapes similar?

If you divide each shape in half, what new shapes can you make?

Can you draw a different pattern using hexagons and triangles?

What are some different ways you can find the total number of apples?

You want to make apple pie.
How many apples will you use?
How many apples will be left over?

What shape does the apple remind you of?
What else has a shape similar to an apple?

You want to organize these items into groups.

How can you sort and group the items so that the similar items are together?

Can you think of a different way to sort and group similar items?

How many bicycles are there?
How many wheels does each bicycle have?
How many wheels is that in all?

About how many more bicycles can you fit on the bike rack? How many wheels would all of those bicycles have?

Which bicycle would a tall person ride?
Which bicycle would a shorter person ride?

What pattern do you see? If this pattern continued, what would come next?

Can you make this same type of pattern using other colors? Can you make this same type of pattern using shapes?

What shapes do you see?
What can you tell me about these shapes?

218

What looks similar in this photo?
If you folded this photo in half, what would you notice?

Find the number in the photo.
If you rearrange the digits in that number, what is the smallest number you can make?

If you rearrange the digits in that number, what is the largest number you can make?

How many bird feeders are there?
Can you think of another way to find the total number of bird feeders?

Are there more bird feeders or birds?

How many more birds would need to join so that there are the same number of birds and bird feeders?

7420
7410
7420
7410

What pattern do you see in this photo?

Can you make the same pattern using shapes?

Can you make a different pattern using the colors black and white?

What is in-between the aloe plant and the tissue box?

What is behind the giraffe?

What is underneath the dinosaur?

What is to the right of the plant?

Is there an even or an odd number of lemons? Is there an even or an odd number of kiwis?

Are there more kiwis or lemons? How many more?

How many kiwis would you need to add so that the number of lemons and kiwis are the same?

225

Estimate the number of rectangles.
Now count the number of rectangles.
How close was your estimate?

If you cut a rectangle evenly in half, what shape can you make? If you cut a rectangle into four even parts, what shape can you make?

Find the number in the photo. Can you rearrange its digits to make a larger number?

Which group has more lily pads?
Which group has fewer lily pads?

Let's make an estimate: About how many lily pads are in the larger group? About how many lily pads in the smaller group? What is the difference between the two groups?

Where do you see reflections in the photo?

Which spice shaker is second in line?
Which spice shaker is first in line?
Which spice shaker is third in line?

Which spice shaker is the tallest?
Which spice shaker is the shortest?
Which spice shaker is the widest?
Which spice shaker is the thinnest?

Which container do you think can hold the most amount of spices? Why?

How many shopping carts are in the first column? How many are in the second column? How many is that altogether?

If half of the shopping carts are used, how many will be left?

If six people return shopping carts, how many will there be?

= WATERING =
CANS

How can you sort and group the items so that the similar items are together?

Can you think of a different way to sort and group similar items?

If you had a small plant, which pot would you use? If you had a medium sized plant, which pot would you use? If you had a large plant, which pot would you use?

What pattern do you see? If this pattern continued, what would come next?

Can you make a different type of pattern using these colors?

What shape do you see?
What do you know about this shape?

How many stools are there? How many people can sit so that everyone has a seat?

If two people are sitting, how many empty stools will there be?

If you doubled the number of stools, how many people could sit?

What shapes do you see?
How are the shapes different?
How are the shapes similar?

Which shape has the greatest number of sides? Which shape has the fewest number of sides?

How many sides does the arrow have? Can you draw another shape with that many sides?

Made in United States
Orlando, FL
04 November 2026

90205471R00026